Ulster in the 19...

Photos from the UTA Archive 2

Norman Johnston

Ulster in the 1950s

All rights reserved. No part of this publication may be reproduced, stored in a retrieval system or transmitted in any form or by any means, electronic, mechanical, photocopying, scanning, recording or otherwise, without the prior written permission of the copyright owners and publisher of this book.

First Edition
First Impression

© Norman Johnston and Colourpoint Books 2008

Designed by: Colourpoint Books, Newtownards
Printed by: GPS Colour Graphics Ltd

ISBN 978 1 906578 13 8

Colourpoint Books
Colourpoint House
Jubilee Business Park
21 Jubilee Road
Newtownards
County Down
Northern Ireland
BT23 4YH
Tel: 028 9182 0505
Fax: 028 9182 1900
E-mail: info@colourpoint.co.uk
Web-site: www.colourpoint.co.uk

Norman Johnston is a native of Northern Ireland. He has written several history textbooks for schools but his first love has always been transport and vehicles of all descriptions. He has been the author of many books on transport subjects, ranging from *The Fintona Horse Tram* to *Austerity Ulster 1947–51*, the first album in this series. Since taking early retirement from teaching he has become a publisher and editor and was a founder of Colourpoint Books in 1993.

Cover photographs

Front cover: Donegall Pass in May 1953. Passers-by pause to watch a parade of vehicles in connection with Road Safety Week.

Rear cover

Left: A railway crane unloads girders for a new bridge over the Galgorm Road at Ballymena in 1957.

Right: Holiday makers at Portrush relax around the Arcadia café in the summer of 1952.

Introduction

This is the second of a series of albums based on the photographs commissioned by the Ulster Transport Authority in the 1950s. Although most of these albums will focus on the transport activities of the UTA, I have widened the interest in this album to look at other aspects of life in the 1950s.

It is hard to beat first-hand knowledge when it comes to writing about something and, as a child of the 1950s, I can at least claim to have "been there and done that". That will immediately date me for most of my readers, but I suspect many of you have picked up this book because you too are heading for 60, if not already there!

"What were the fifties like?" younger readers will ask. As Northern Ireland emerged from the austerity years that followed World War Two, the pace of change was still relatively slow, though it began to speed up from about 1956. With hindsight, I now realise that life in the fifties was the last gasp of what was really a thirties lifestyle, albeit interrupted by war. By that, I mean that someone from the 1930s, transported suddenly into the 1950s, would have found a lot that was familiar, including social attitudes and morals. In contrast, if someone living in the 1950s had been suddenly transported into the 1970s they would have found a lot that was strange and, indeed, threatening, especially in the Northern Ireland of the 1970s.

A study of seaside photographs of the fifties gives the impression that people of that era were stiff and formal. Seaside visitors tended to wear suits and jackets, and even hats. However, this impression is false and can be easily explained. Times were much more frugal and only the well-off had what today we would call 'leisure-wear'. Most people in the 1950s had only two types of clothes – their 'working' clothes and their 'good' clothes. 'Good' clothes were worn on Sundays, on special occasions like parties, going out and days off. Thus, in wearing your 'good' clothes to the seaside, you were really saying "I am on holiday, I am relaxing. This is my day off". Remember, that 'working clothes' did not just mean what manual workers wore, it also applied to shop assistants and office workers, or anyone who normally wore a uniform, like a postman or bus conductor. For office workers, normal attire was probably a tweed jacket, but one that was a bit worn and probably reinforced at the elbows and cuffs by leather patches. On their day off, the 'good jacket' came out.

Although not illustrated in these photos, toys in the fifties were an interesting mix of old and new technology. Twenties- and thirties-style tinplate toys were still popular – remember those tinplate buses with tin-printed passengers depicted at the windows? Diecast Dinky toys were in their heyday, still with crude wheels and no glazing or interiors. From 1957 on, they were challenged by the more sophisticated products from Corgi and, from 1959, by our own 'Spot On' models, manufactured off the Castlereagh Road in Belfast. However, at the other end of the scale, plastic was the up and coming material for toys. Early fifties plastic toys were very brittle and smashed to bits if you stood on them but from the mid-fifties softer plastic appeared, especially for those plastic soldiers and cowboys that fifties schoolboys loved. Plastic dolls for girls saved many a tear when a favourite was dropped accidentally.

For me, 1957 seems to have been a watershed, between the old ways and the new, in the pace of change. It was the year when large swathes of the railway system were closed, with everything in Counties Monaghan, Cavan and Fermanagh, and most of Armagh, closing in one swoop. It was the year when

Elvis became a big name and people started worrying about 'Teddyboys'. It was the year when the first Sputnik (a Russian space satellite) was launched and could be seen over our province with the naked eye at night. Space fiction, like Dan Dare, was no longer just fantasy. More and more people were buying cars and the old pre-war 'upright' cars, seen in these pages, became rarer. There was a bit more money about and the demand for a week at the seaside and daytrips by bus was insatiable.

We were all happy to holiday in Newcastle, Bangor, Bundoran or Portrush. Scotland and England were the normal destinations for the better-off. The concept of regular continental holidays was still at least ten years in the future. Today, you don't think you've been on holiday at all, unless you have been in an aeroplane!

The Unionist regime in Stormont was secure and largely unchallenged in the 1950s. The part played by the province in the war was still gratefully remembered by the British Government, who were little inclined to interfere in the more unpleasant aspects of life here. However, the development of free secondary education, and the opportunities for patronage created by the expansion of modern public housing, led to tensions that would erupt in the late 1960s, as a generation went through school who were of a mind to challenge the status quo and the discrimination prevalent in society here.

Turning to another aspect of life in the fifties, the thatched houses that feature in some photographs remind us that in this era such traditional dwellings, which now feature only in the Ulster Folk and Transport Museum and the Ulster-American Folk Park, were the normal housing for so many of us. I remember noticing in the fifties, as I was growing up, that about one house in three was thatched and another third were slated, as opposed to having modern tiled roofs. About half of all houses lacked flushing toilets or running water.

Although a 'townie' these days, fifty years ago I was well used to staying with aunts and uncles in Fermanagh who had no electricity, no inside toilets and who had to draw water from a well some distance from the house. To this day, I could still light an oil lamp if I had to and wax eloquent about the power of a 'Aladdin' lamp. Like many readers, I remember free range poultry and hand milking and in the summer as a teenager took my turn at 'lapping' hay. Oh for the taste of a boiled egg from those days, compared to the modern version! And, as for the flavour of real bacon and proper farm-made butter, there is nothing today to compare with what we took for granted in the 1950s.

I remember being at an aunt's house with outside privy and no electricity in 1959 when an elderly Canadian cousin and his wife paid a visit. After a short time, the cousin politely asked to be shown the 'rest room'. We were all puzzled and my aunt showed him into the bedroom, assuming he wanted to lie down! When the penny finally dropped, he was shown up the yard to a small wooden hut with a six foot drop to the cess pit. He came back somewhat subdued and his wife didn't dare to ask.

As these pages will show, people in the fifties were eager to soak up any entertainment that offered itself, whether it be a circus, a Road Safety Week parade, a bus trip, or the Pierrots at Newcastle. It is sobering to remember that television did not come to Ulster until 1953, when national BBC transmissions began in time for the Coronation. Even then, it was another six years before UTV appeared, so the old cry of "What's on the other side?" did not begin until 1959! Our diet of afternoon television in the late 1950s and early 1960s revolved round cowboy films, and cartoons like Topcat, Yogi Bear and Micky Mouse. Our role models were Davy Crocket, Roy Rodgers and Tonto. No video games for us – it was real action, sixguns at the ready, quick on the draw, "Dar-dar, you're dead". What a time we had!

Welcome to the 1950s. Okay, there was a lot wrong with it but, somehow, if I could create my own Heaven, it would involve half cab buses, steam trains, Aladdin lamps, the smell of smouldering turf and, above all, real bacon!

Beside the Seaside

We start this book with an almost timeless scene at Carrickfergus Castle on 27 July 1952. Yet, although the castle itself is little changed, apart from restoration work on the tower since then, there are 1950s clues in the boats and the cars in the carpark. Today the carpark would be much busier, of course. The cars seen in 1952 seem to be mostly Morris products, but the row end-on to the camera includes a Ford V8 and a Riley, whilst behind the latter is a pre-war Hillman Minx. Note the Landrover to the left, in front of which is a Ford Anglia.

No doubt to help illustrate bus tour brochures and the like, the intrepid photographer was dispatched to Portrush on 31st August 1952. His efforts at the Arcadia tell us quite a bit about the Ulster seaside in the early 1950s.

Whilst there have been changes in half a century to buildings and the layout of the lawns, the basic custom of sitting on a bench seat to sunbathe and watch the world go by is still with us! What has changed since the 1950s is how we dress. In 1952 most people 'dressed up', rather than 'down' for a day at the coast. No denims or even tee-shirts in 1952! Sunday suits and overcoats across the arm were the norm, though few hats are in evidence.

The tide is fairly well in as the family of three stroll along the top path towards the Arcadia café. The boy on the right carries a bucket and has typical 1950s short trousers with cross-strap braces. On a personal note, I was about the age of the boy here when I went on my first family holiday to Portrush in August 1953. In the days when people came to Portrush by train or bus, there were no cars to conveniently leave your coat in, so day trippers were stuck with them, even when the weather turned out good. Although not obvious in this picture, the famous Arcadia Ballroom had not been built in 1952. It was to the right of the café. It has come and gone and the building once more looks much as it did in 1952.

Portrush bus station in 1952, seen from Barry's Amusements. Nowadays, the bus station is much further from the sea front. Portrush was characterised by a high proportion of double-deckers, an advantage of the area being that the railway bridges were quite high and, of course, with heavy traffic, 48 seats was an advantage. The bus on the right is a 1936 Leyland TD4, whilst in the foreground is a wartime TD7.

Viewed from the opposite perspective, we look towards Barry's, with the 'big wheel' and helter-skelter in evidence. Three single-deck postwar Leyland PS1s add variety to the bus scene. On the left we see the clock tower of the railway station, still a landmark in Portrush. On the right, between the fences and to the right of the buildings, can be seen the path leading towards Barry's other attractions, like the dodgems and the ghost train. Below the fence, hidden from view to the right, was my favourite attraction, the Peter Pan railway, basically like the ghost train design but outdoors and without the scarry bits! Many a happy hour was spent here in 1953, driving round the gardens in engines like *Snow White* and *Aladdin*, vigorously ringing the bell, and getting on and off at the wooden Tudor style station, the latter new in 1953.

This interesting view of Portrush was taken from Castle Erin holiday home in May 1954 to illustrate the effects of coastal erosion. Barry's is the white building on the right, with the dodgems, ghost train, mirror maze and other features. In the foreground is the path from Castle Erin and between it and Barry's is the harbour branch of the railway, truncated beyond Barry's and used for storing carriages on busy Saturdays. Just below the white terrace in the centre of the picture is the Peter Pan railway, its Tudor style station visible to the left of the cluster of buildings alongside Barry's. On the left is the West Strand, popular in the afternoons, when the sun was on this side of the town. The town towers above the beach, the upper row being the back of Main Street. Portrush is dominated by its guest houses and hotels, including the UTA-owned Northern Counties, visible on the left horizon. At the top right is Portrush Town Hall, the venue for summer theatre. Castle Erin was demolished in 2006.

A tranquil view of Ardglass, Co Down in August 1952.
In the 1950s, Ardglass had a population of about 750. Taken from the harbour, the picture shows a typical seaside Ulster town of the period. The skyline is dominated by churches and Jordan's Castle, named after Jordan de Saukeville, a Norman knight, who hosted King John on a visit to the town in the early 1200s. Ardglass was famous for its herring fishing and two of the small fishing boats of this era are moored.
The gulls are spaced out as if they are each occupying their own berth!

One of my vivid memories from the 1950s was of a flat bed lorry, such as the Leyland Comet in the background here, going round our estate in Portadown with the driver shouting what sounded like "Her'ins alive!". It would be stacked with barrels of herrings, like those in the background, and half a dozen of these fried provided an economical and tasty meal for three which was regarded as a special treat. Here the fisherwomen of Ardglass sort the catch before loading into the barrels. The customer did the gutting! Potted her'ins were also a popular meal. Gutted herrings were covered in flour, rolled up, tied with thread, dipped in vinegar and boiled. Note the then-current type of Hillman Minx parked on the quay.

Two Ardglass lads wander round the harbour and contemplate the fishing boats in another scene from 1952. The names are carried on the sterns, BA 274 being *Queen Victoria*. A mid-1930s Austin Seven can just be seen in the right background and in the distance some visitors are exploring the harbour.

On the streets

1953 was Coronation year and, on 2 May, Road Safety Week was marked by a parade of vehicles in Belfast. Preparing to set off at Duncrue Street we have some interesting vehicles. The 1950 UTA Commer articulated lorry, with its prominent fleet number, carries a racing car of the period, complete with driver (possibly here to compete in the Ulster TT Race at Dundrod), accompanied by an excited Boy Scout guard of honour (My, my! What would 'Health and Safety' think of this today?). Behind the lorry is a new Lagonda drop-head coupé, DZ 5779, with another contemporary sports car behind, possibly a Daimler Straight Six.

The buses in the parade turn into High Street from Victoria Street with the Albert Clock (Belfast's 'leaning tower of Pisa') in the background. The Custom House is above the second bus. At first I was puzzled by the inclusion of such an old bus in the parade, although it had been specially repainted for the occasion. However, both buses have windows posters and, from another photo, these read "New, Coronation Year 1937" and "New, Coronation Year 1953". EZ 1826 was a 1937 Dennis Lancet II and some of these still worked into Portadown from Lurgan in the late 1950s, during the busy summer season, when all sorts of old crocks in the fleet were pressed into service, as well as hired-in Belfast Corporation buses.

The newer vehicle of the pair on the previous page, OZ 833, was a new Royal Tiger front entrance bus with 'dual purpose' seating. These had coach-type high-back seating for Private Hire work and the UTA were very proud of them, but they had limited leg room, so they were not really full coaches – though that was as good as it got on the UTA! The two tone green livery was attractive and I have fond memories of trips from Gilford to Newcastle on these in the mid-1950s. As can be seen, any special event attracted plenty of spectators in 1953, though I dare say the racing car was more of a highlight than a bus! In 1953 there was still two way traffic round the City Hall. Everyone is well 'happed up' for the changeable Ulster weather typical of May.

This is another view of the 2 May parade and is also our cover photo. Having gone round the City Hall, the probable route was Bedford Street, Shaftesbury Square and now the vehicles are going down Donegall Pass towards the Ormeau Road, with the Gasworks in the background. The main interest in this shot is provided by the passers-by. All categories of 1950s fashions can be seen, along with two prams whose occupants are now 55! The UTA had included two of its latest double deckers, the long white poster on the nearest being a 1953 version of the later 'Green Cross Code'. Amazingly, 'Fast Taxis' are still with us, still using a version of the same telephone number, but now based on the Newtownards Road!

On the Buses

Opposite: A scene at North Street bus yard, Belfast, on 6 September 1951, showing a conductor sorting parcels and mailbags into the rear luggage locker of a front entrance bus. These buses still had a roof rack, as can be seen from the ladder. However, the rear locker was a useful advance on the half-cab buses seen in the background.

The picture is a reminder of the heavy parcels and sundries traffic carried by bus in the 1950s. In the days before Sellotape, all parcels were done up with string. Note the leather cuffs on the conductor's tunic.

This page: This picture was one of a series of photographs taken at Smithfield on 6 June 1952 to illustrate the uniforms worn by busmen, in this case by a Driver and an Inspector. The tunics are both double-breasted but the driver's has leather cuffs to take account of contact with the steering wheel. All busmen carried a cap badge indicating their role – 'Driver', 'Conductor', etc. Trousers are turned up in 1950s style.

The bus is MZ 7786 (D924), a 1950 Leyland PD2/10 double decker. The prefix to the fleet number on UTA vehicles indicated the year of the chassis – D for 1950, E for 1951, etc. A previous series of these letters had ended in 1946 with Z. Note the opening driver's window and the fog light. The white box on the nearside could be illuminated to indicate 'Bus full' to waiting passengers.

There is something vaguely comical about this picture of the new rail/road terminal at Ballymena on 21 July 1952. I think it is to do with the way the layout of the waiting/boarding area resembles a railway station!

Every bus in sight is one of the standard PS1 single deckers of 1946–48, of which the UTA had 410, making them the largest single operator of this type in these islands.

In the background is the bus garage and Ballymena railway station is conveniently to the right, out of shot.

The girl looking at the camera is a reminder of an age when many bus travellers had suitcases with them.

The skid pan at Duncrue Street, on 16 September 1954, is reminiscent of a famous scene in the Reg Varney comedy film *On the Buses*. Part of the training of a new driver was to give them practice at dealing with a skid. An area of tarmac was continually hosed and the driver raced at the wet area and slammed on his brakes, sending the bus into a spin. In this skid, the driver is losing his rear, so he is correctly steering right to regain control. Needless to say the buses used for this were not the 'new kids on the block' and the vehicle here was one of a number of ex-Royal Army Service Corps buses brought over from England in 1942–3. Fleet number R660 was a 1938 Leyland TS8, rebodied at Dunmurry in 1946. Just in case, the area in the background is conveniently lined with all the old wrecks of the day – lorries from the late 1930s – most of which are heading for the breakers yard anyway.

On 27 July 1952, the photographer visited Kilroot to photograph a cottage which was just beside the railway station and, sadly, is now gone, destroyed by vandals in the 1980s. Thatched houses were common in the 1950s, my personal recollection being that about a third of rural houses were thatched and mostly white-washed, as here. This house has a hip roof, which was less common in the east than it was in Fermanagh and Donegal and was oval in shape. The famous Jonathan Swift lived in it when he was Minister at Kilroot in 1694 and said the devil would never catch him in a corner!

County Antrim

Whitehead, seen here on 23 September 1952, has expanded rapidly in recent decades. Today, much of the high ground spreading towards Islandmagee, in the background, is covered by houses. The photographer's vantage point is from the main A2 road towards Carrickfergus and can still provide this panorama today. The telegraph poles on the right mark the railway line, and the promenade and coastal path to Blackhead, both built by the BNCR, are clearly visible. When buses first served the A2 coast road in the 1920s, the good citizens of Whitehead would not permit them into the town so, instead, they terminated at Wisnom's corner, a stage stop still shown in early Ulsterbus timetables for the 163 route.

Beyond Black Head, the BNCR built a spectacular path along the face of the Gobbin's Cliffs on Islandmagee in the 1890s. The path had deteriorated quite a bit during the War years and a half-hearted attempt had been made by the UTA to restore some of it in 1949. One of the more spectacular features of the path was this tubular girder bridge, which carried the path over to the rocky sea stack in the background and, by another bridge, back to the mainland. When this photograph was taken on 27 July 1952 the path was still just about passable, but in the intervening 56 years most of the bridges have gone. However, recently, cross-border funding has been agreed to restore about half of the path to its former glory.

Brown's Bay on 27 July 1952. This is a picturesque spot, though not the best known of Ulster's beaches. Situated on the northern end of Islandmagee, it is familiar to those living in the Whitehead/Larne area. Note the bathing cap on the lady in the centre, a very common accessory in the 1950s to protect perms. Once again, a man walking the beach in a suit was par for the course in 1952.

The village of Glynn, Co Antrim, on 23 September 1952 is the setting for this pleasing cameo of rural life. The thatched houses in the background are gable-ended and more typical of nineteenth century housing in the east than the earlier view at Kilroot. They represent the 'long house' style, with a series of parallel rooms linked by a long corridor at the front running the length of the house, Expansion was by simply adding a new room at one end. The lady with the pail reminds us that many households still had to draw their water from a well in the 1950s and even into the 1960s.

A journey up the Antrim Coast Road from Larne, on the same day, brought the photographer to Carnlough. Here, on a perfect afternoon, we have the spectacular view of Carnlough bay and the village, from the limestone quarries in the hills above. Across the bay, we see Straidkilly Point and, beyond it, Park Head. Between them, hidden from view at the inlet, is the village of Glenarm. Glencloy, another of the nine glens, lies between the two sets of hills on the right.

The lime kilns at Carnlough in August 1951. Visible on the hills is Creggan quarry which, with Gortin Quarry out of view to the left, employed so many in the Carnlough area. A rope-worked railway ran from left to right across the picture – its embankment can be seen to the left of the kilns. This linked the quarries to the harbour at Carnlough and crossed the Coast Road in the village by the stone bridge that still exists. The winding house for the railway can be seen protruding above the stone heap to the right. By 1951 the kilns were disused, as most of the processing was by then done by a crushing plant up at the quarries and a whiting mill further down the line. The lime powder was mostly exported to Scotland and England for industrial use and for road construction.

The beach at Ballycastle, adjacent to the harbour wall, had plenty of activity going on when this scene was recorded in August 1957. Children predominate, watched by their mothers or aunts. The well-known landmark of Fair Head dominates the background and the west coast of the Mull of Kintyre, in Scotland, can be discerned in the distance. The rowing boats, with their prominent 'CE' registrations, add interest. Some carry names, like CE112 *King George VI* at the water's edge.

On 6 October 1952, the photographer was sent on assignment to Derry to record bus loading points in the city. The one seen here, outside the Guildhall, is still a loading point today, but for city sightseeing tours, rather than service buses. The Northern Bank in the background reflects the solid image of such institutions. The UTA bus on the 'B' city service to Glendermott Road is a 1949 Leyland PS2. These single deckers resembled the PS1s externally but had the larger Leyland '600' engine which made them an ideal choice for Derry, with its steep hills. Many were later rebuilt as double-deckers.

This view, on 31 January 1952, shows a room in the City Hotel, Londonderry set out for a dinner, quite possibly for some sort of UTA function. Although the UTA owned several former railway hotels, this was not one of them. The picture shows how an hotel maid of the period dressed. The tapestry in the background is most impressive and seems to be a composite of several Northern Ireland tourist icons, including Dunluce Castle, the Watergate in Enniskillen and Errigal Mountain.

Another popular Ulster holiday resort in the 1950s was Newcastle, Co Down. I first made its acquaintance in July 1954 when I lived in Gilford and it became the obvious destination for our annual church excursion by bus, using one of the light green Royal Tigers depicted on page 16. Two years later I enjoyed a family holiday here and a daily ritual was to come to the bandstand, seen here in August 1952, to watch the famous 'Pierrots', as we termed the group of entertainers who did puppet shows and the like. In this view, with Slieve Commedagh and Slieve Donard in the background, the crowd are gathering for the show. The notice indicates that adults were charged a shilling, and children sixpence.

At Newcastle, the bus garage was adjacent to the railway station, then still open for GNR trains, and in the background in this May 1952 shot. The actual purpose of the photograph was to record a UTA poster site, this one aimed at encouraging private hire work. Of interest are the old fashioned tall petrol pumps, with their bottles of engine oil. The black car is a pre-war Austin Seven 'Ruby' and the cream one a post-war Austin Devon, with what looks to be a Ford Eight beyond.

Market day at Ballynahinch in August 1952 underlines the fact that horses still played a big part in local transport in the 1950s. Facing the camera, with the white horse, is an Irish sidecar, once common everywhere but now rarely seen outside Co Kerry. Sideways to the camera, in front of the Market House, is a 'Scotch' cart, used by many farmers to carry virtually everything from groceries to milk churns. The elderly bus is a 1936 Dennis Lancet with Cowieson body. It was withdrawn in 1956. The sporty looking car on the right is a Riley RMA, a 1.5 litre car produced between 1946 and 1955.

A view of the railway bridges across the River Bann and the Toome Canal at Toome, Co Antrim, on 23 September 1952. The picture is taken looking north from what is now Canal Walk, but was then the tow path. The railway was still open for goods traffic at this date, but closed in 1959. For a period part of the line (on the east side of the river) was converted into a road, as part of a one-way system through Toome, and the bridge was demolished after the Toome by-pass opened in 2004. Note the swing bridge over the canal.

We now move south to Armagh city for two photographs taken on 10 May 1952. The first is a rather artistic composition with the Catholic cathedral framed in an archway. The album does not specify the location or purpose of this shot and clearly the houses in the foreground are derelict and very ancient.
From the angle of view I would guess that we might be looking at St Catherine's Court before it was redeveloped. This is a dead end street with a courtyard at the white house. Below the cathedral, the terrace houses are on Cathedral Road and a corner of Edward Street is also visible in the distant right.

There is no problem with identifying this one, however! Here we are standing at the bottom of College Hill, facing the side of the Courthouse, with The Mall to the left. On the horizon is the Church of Ireland cathedral, much less imposing than its Catholic neighbour. Note the bus parked at the traditional departure point on College Street. The cars, even those on College Street, are all fairly old. In the foreground is a pre-1938 Morris Eight, facing what looks like a Riley Nine saloon. About to turn into the Mall is an Austin, most likely a pre-war 'Ten' Cambridge, though from this distance it could be any of the 1936–40 Austin family with their characteristic twin rear windows. Note the pony and trap emerging from The Mall.

In the last week of January and the beginning of February 1953 a Co-Operative Travel Exhibition was held in Belfast, probably in the upstairs of the Co-Op building. Naturally, this featured a UTA display, which is on the right, and featured the miniature double-deck bus illustrated in *Austerity Ulster*, the first volume in this series. It also offered caravan holidays, a possible reference to the converted railway coaches rented out to holiday makers at Ballycastle and other places.

The British Railways stand included models of 'Jubilee' and 'Royal Scot' class locomotives, and one of the 1928 LMS 'Duke' class ships. BR was promoting day trips aimed at football fans to Bolton, Burnley, Preston and Manchester.

Note the Pan American stand on the left.

This is the UTA display at the Balmoral Show on 26–29 May 1954. It has to be said that this was not a very exciting or ambitious stand, the only thing promoted being 'coach tours', slightly ironic, in that the UTA didn't actually possess proper coaches! (see page 16) One significant change from stands illustrated in *Austerity Ulster* is the inclusion of the Great Northern Railway Board which is advertising 'Diesel Expresses'. The GNR had been bought out by the two governments in Ireland in 1953, and was run by a joint Board, in a unique arrangement that lasted until 1958 – arguably the first 'cross-border body'. In the right background Aga cookers are being advertised, one of the few products in this picture still with us.

I have not included many train photographs in this album, as I intend to do a railway-themed album later. However, I could not resist this shot of a brand new diesel train being wheeled out of Duncrue Street onto the traverser on 29 December 1952. Following the successful introduction of a three-car AEC-engined train in 1951, the production version in 1952–54 used Leyland engines that were inter-changeable with the buses. Fourteen three-car sets were produced and known as Multi-Engined Diesels, or MEDs. With pneumatic doors and (with some exceptions) all-metal construction, they were far ahead of anything available in England at that time and, indeed, for some time to come.

The lack of proper coaches in the UTA fleet (see page 16) created a problem in 1951 when Southdown, a bus and coach company based in the south of England, decided to enter to enter the Northern Ireland travel market with extended tours from England. Because the UTA had a legal monopoly for bus services in the province (outside Belfast) Southdown could not legally operate their own luxury coaches – but the UTA could not provide coaches! So a legal compromise was reached – Southdown sold two of their own coaches to the UTA, which added its own fleet numbers and insignia. The price would be refunded on return, less depreciation. A newly delivered Harrington-bodied Royal Tiger is seen at Duncrue Street on 4 May 1956, lettered UTA, but wearing the pleasing Southdown light green livery.

In contrast to the new diesel trains described on page 40, and which took over all services on the Bangor line, something very much more ancient is seen at Queen's Quay on 31 January 1955. This old carriage is Dublin Wicklow and Wexford Railway 3rd Class coach No 48, dating from 1835, which had been donated by CIÉ for the new Belfast Transport Museum. Initially, until the museum opened at Withim Street in 1962, the exhibits were displayed in the old railmotor shed at Queen's Quay (in the right background). In this scene, railwaymen and enthusiasts, including John McGuigan on the left, are helping wheel the old coach towards the points for the adjacent line, from where it was shunted in by a railcar. The ex-BCDR steam engines in the background, awaiting the cutter's torch, are 4-4-2T No 219 and one of the big Baltic tanks.

This is a view of Lisburn railway station on 3 November 1959, about a year after the UTA had taken over the northern portion of the GNRB. Not a lot has changed here in forty years, except that the trees in the centre have given way to a station carpark. If anything, thanks to NIR, the station is in better shape now than it was then. The lorry parked on the left appears to be a Post Office Telegraph one, the type used for the erection of new telephone lines, with ladders carried on the side and telegraph poles inside. The nearest car is a post-1948 Ford Prefect, with two Vauxhall J Type taxis among the other vehicles parked outside the station.

The closure of some railway lines in 1955–56 and the raising of bridges on others, enabled the UTA to extend the field of operation of its double-deckers. 158 of the Leyland '600'-engined PS2 single-deck buses were rebodied in 1956–58 as high bridge double-deckers. This shot shows one such conversion under way at Duncrue Street on 28 June 1957, with a near-completed example in the background. Still bearing its old registration, this particular vehicle became UZ 7733 in its new guise. Unlike the earlier buses, these vehicles had metal body frames by Metro Cammell Coach Works (MCCW).

All double-deckers had to undergo a tilt test before being cleared for service. Here, the second vehicle converted, UZ 682, passes the 28° legal minimum at Duncrue Street on 7 May 1956, watched from a safe distance by some senior staff. The destination blind will need sorted out.

The completed product is illustrated by this view of UZ 7726 on 28 June 1957. The destination blind should read 'Templepatrick' but is not yet properly set. Unlike the open platforms of Belfast Corporation double-deckers, the UTA ones had folding doors, which were much appreciated by lower deck passengers in the winter. These vehicles looked more modern than the earlier low-floor types and remained in traffic into Ulsterbus days, the last going in 1972. They seated 60 and had very lively springing, as many upper deck passengers could testify!

The very first of the high-bridge double-deckers was proudly displayed, in UTA two-tone green, in the parade for the Lord Mayor's Show in Belfast, on 26 May 1956. This shot was taken near the Shaftesbury Square end of Great Victoria Street, the parade then turning left into Donegall Pass. Behind the bus is one of Cowan's Ford Thames flatbed lorries, advertising the Heysham route, followed by a Bedford advertising the Anglo-Continental container service.

Compared to NIR and Translink in more recent years, the UTA undertook few major infrastructure projects in the 1950s, mostly new bus stations and improvements to Belfast railway termini. However, in the late 1950s some railway bridges were widened and raised, notably at Derriaghy, Templepatrick and Ballymena. This is a record shot of the GNR railway bridge at Derriaghy, on 15 June 1956, prior to its replacement the following year. The new bridge was a steel plate-girder bridge, formerly in use on the BCDR main line just west of Comber. This view is looking south west, with Gallagher's 'Blues' and Exide batteries advertisements prominent. Perhaps Girling brakes would have been more appropriate here!

Concrete combined bus posts and signs had been introduced by the LMS NCC in 1932 and a few were erected by the NIRTB, the immediate predecessor of the UTA. In the 1950s the UTA began to erect them in large numbers and the photographer was sent out to record some in October 1957. Most said 'Bus Stop' or 'UTA Stage' but this more unusual example has 'Bus Stage', quite where I am not sure, but possibly between Dunmurry and Derriaghy. The photographer's car is a 1949 Vauxhall J Type.

Antrim in October 1956. This was one of a number of photographs taken of road passenger sites that were identified for the erection of electric clocks, presumably because they were adjacent to bus departure points. However, the focal point of the picture is Fred Young's garage and illustrates the restricted sites that many street-side garages occupied at this time. The garage part of the premises is down a narrow entry in what is presumably a workshop in the yard. The shop sells a variety of parts and accessories, including Shell and Vacuum oil and the petrol pump is also Shell. Older motorists will remember that, up to the 1970s, petrol pumps were attended. You didn't even need to get out to pay!

One of the biggest jobs undertaken by the UTA at this time was the reconstruction of Galgorm Road railway bridge, in Ballymena. This was the bridge adjacent to the railway station and the whole process was extensively photographed for the UTA archives. The work began early in 1957 and this view was taken on 13 March 1957, looking east with the station on the left. Already the piers for the new wider bridge are taking shape. The original bridge carried three tracks at one time as, until 1933, the narrow gauge to Larne occupied the side away from the camera. The 'Road Up' sign on the left was a common sight in the 1950s. How many young people today would appreciate this message?

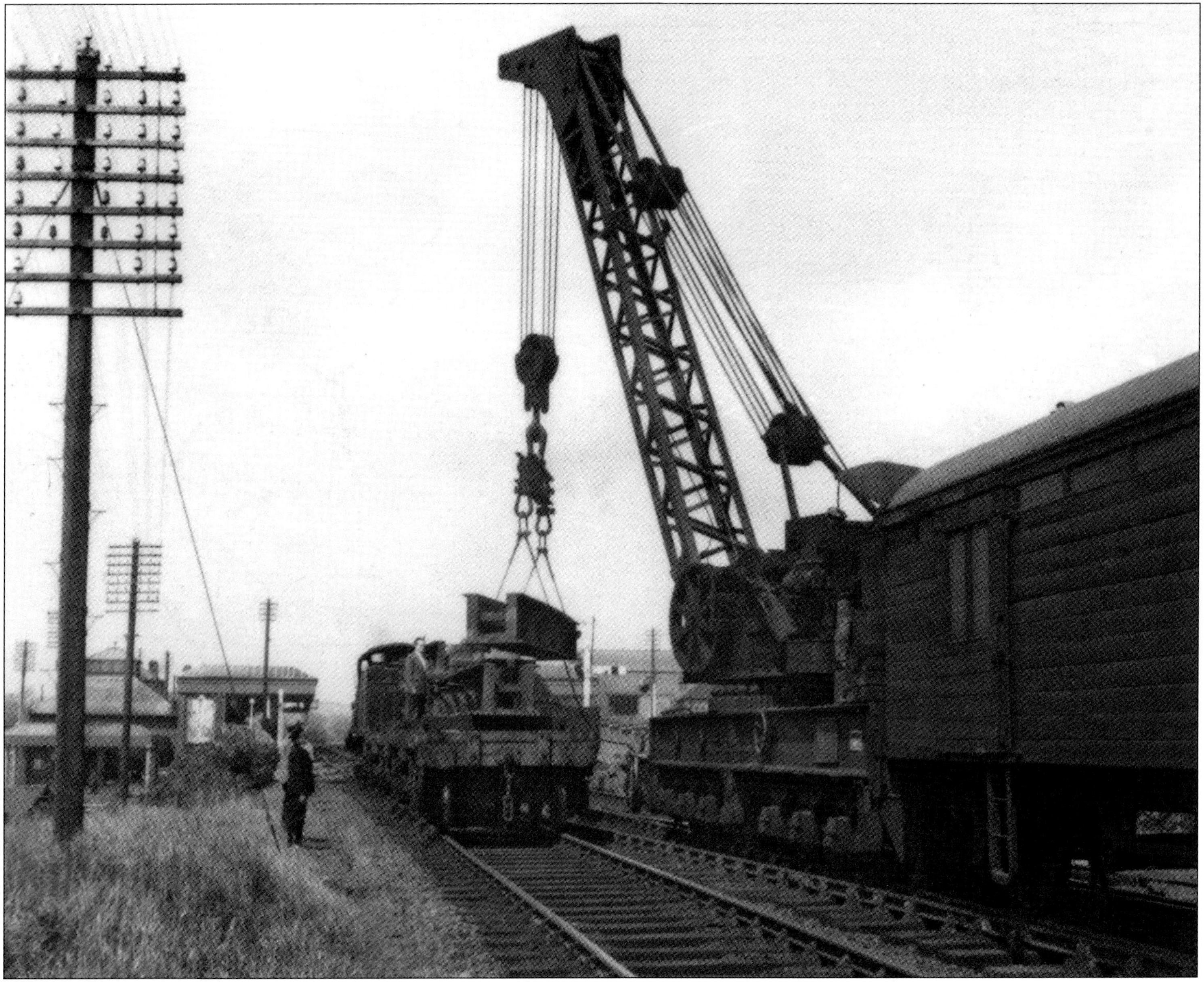

These two views show different stages in the reconstruction of the Galgorm Road bridge, Ballymena in 1957. Opposite, on 22 May, the NCC 36 ton railway-mounted steam crane is unloading the under-girders, which will support the trackbed. The locomotive is 4-4-0 No 84. On this page, looking towards Belfast on 10 July 1957, the first of three large girders is being placed. Two of these would form the sides of the bridge, and the third was placed between the running lines. They had a double function as they also braced the load-bearing under-girders seen opposite. Note the absence of hard hats on workmen of this era. When the crane was on the move the crew and their kit travelled in the two converted brown vans.

Oxford Street

One of the biggest UTA infra-structure projects in the late 1950s was the construction of the new Oxford Street bus station in 1960. Work on this began in late 1959 and these next few pictures record the area as it was on 20 October 1959, when work commenced. On the left we see the corner of Chichester Street Fire Station, now occupied by Laganside Courts. In the foreground is Oxford Street and, behind the hoarding on the right, is Laganbank Road. Today this whole area is vastly changed in character. The site of John Arnott's is now the Riverfront office block, featuring the Harbour View oriental restaurant, and the docks in the far background look very different today. However, the sewage pumping station in the bottom right corner is still there, as is the Queen's Bridge in the middle distance.

This view is taken from the same vantage point but the photographer has swung the camera to the right to record the site clearance underway for the bus station. Today this area is occupied by the Waterfront Hall, which opened in 1997, and the Hilton Hotel.

Note the line of ten double-deck buses parked along the Laganbank Road, adjacent to the goods-only railway line which linked the Great Northern Railway to the NCC via a tunnel under the Queen's Bridge. The girder bridge in the background linked the GNR with the former BCDR line to Bangor. In the 1960s this viaduct was known popularly as the 'Shakey Bridge' and was replaced in 1976 by the present Lagan Viaduct.

The carpark in the foreground will be of interest to car enthusiasts, containing a wide variety of models, including such 1950s icons as a Riley Pathfinder and an MG Magnette.

Another view, also taken on 20 October 1959, shows the junction of Laganbank Road (left) and Oxford Street (right), looking towards the Royal Courts of Justice (to the right of the Silver Cabs ad) from which the two previous pictures were taken and the distant Gasworks on the Ormeau Road. The railway line referred to in the previous captions is behind the advertising hoardings and below the Silver Cabs ad is the bridge taking it under the Laganbank Road. The car in the foreground is a Monaghan-registered 1950-56 Ford Consul and the buses are a 1947 PS1 half-cab, a 1957 Leyland Tiger Cub and, in the distance, one of the new 1959 Leyland PD3 front-entrance double-deckers.

The three hoardings in this picture can be seen behind the bus in the picture on page 54. The advertisements are a study in themselves, reflecting a period when ads for Kit-Kat could happily sit beside those for whiskey and one featuring the Chivers 'Gollywog' which would cause such offence today. The featured cars are a 1948 Vauxhall L Type Velox, a 1955 E Type Velox, a Series I Land Rover, a 1954 Austin Cambridge and a 1956 Ford Zephyr. The single storey stone building on the right was originally Queen's Bridge station, built by the former Belfast Central Railway, and also its headquarters.

The Circus comes to town

Chipperfields Circus at the Fair Green, Portadown on 2 October 1959. In the 1950s, the arrival in town of a circus usually caused a wave of excitement among children, and parents were only too willing to bring them along. The occasion depicted here was my own first circus and very exciting it was, with performing elephants, trapeze artists and the clowns. Chipperfields had a much bigger tent than the Irish circuses. A Standard Ten Companion can be seen in the foreground.

The UTA's interest in this event was because it involved experiments in the loading of circus animals into trains (possibly unsuccessful). Here two elephants are being led towards a UTA cattle lorry which will convey them to the goods station. The houses in the background are in Jervis Street and the railway line to Dublin is below the wooden fence. The living conditions in the Fair Green must have been far from ideal for the circus folk, with all the mud requiring everyone to be in Wellingtons, but no doubt the hippo was happy!

It was important to ascertain that, once in the cattle truck, an elephant had room to turn and come back out again! The photoshoot at the Fairgreen shows elephants both entering and leaving the lorry. Note the pulley and wires for raising the ramp. Once at nearby Portadown goods yard, the cattle truck was reversed up to the platform and the animals walked across to the waiting goods vans at the opposite side. There is no evidence that Chipperfields' animals actually moved by rail but, in 1964, Bertram Mills Circus was transported around Ireland by train in this fashion.

On 1 October 1958, the Great Northern Railway Board was broken up and its assets divided between the UTA and CIÉ. The UTA found itself with a third Belfast railway terminus to add to York Road and Queen's Quay. Prior to some limited modernisation of Great Victoria Street station in 1960, the existing GNRB terminus was recorded on 5 November 1959. This view shows the barrier to Platforms 4 and 5. Behind the travel posters is one of the GNR's 6-wheel parcels vans that could be found at the rear of many trains. The posters are mostly tourist-related.

This view shows the ticket gates for Platforms 2 and 3. These wrought iron gates and railings were replaced in the modernisation which followed and Platform 5 was removed completely to provide road access to a new bus yard. Note that passengers are encouraged to have tickets *and contracts* ready for inspection. The latter is possibly a reference to the MoD travel warrants used by servicemen/women on leave, who made journeys combining British Railways, a shipping company and the UTA. Details of the Warrant had to be noted at ticket offices so that the cost could be reclaimed from the MoD, the serviceman/woman needing to retain the document for the next stage of his/her journey.

This is a close-up of the area to the right of the previous picture. Note the CIÉ timetable, useful for travellers planning to continue on from Dublin to places like Waterford, Cork and Killarney. Chalk boards were used to give details of short term changes to the normal timetable or to advertise special public excursions. This one, in superb script, tells travellers that the normal 2.15pm connection from Portadown to Dungannon will not run. The ticket collector keeps a wary eye on the photographer.

Few shots of moving trains appear in the UTA archive photographs, but this busy scene at Portadown Goods station on 22 October 1959 is an exception. This area became the site of the present Portadown station in 1970, the passenger station in 1959 being in the far distance, at Watson Street. The two tracks and crossover in the foreground are the entry tracks to the goods yard. On the left a down cattle train is heading for Belfast, the steam from its locomotive being visible above the approaching railcar, which is the 4.25pm to Warrenpoint. On the right another steam engine is shunting in the goods yard and a third is passenger pilot in the main station, half a mile away. Portadown had three goods departures up the Derry Road every night, and three arrivals, so I often heard the distant sound of trains shunting during the night, in my teenage years in Portadown!